SAND

Written by
Noah Leatherland

CREATURES OF THE COAST

BookLife PUBLISHING

©2023
BookLife Publishing Ltd.
King's Lynn, Norfolk
PE30 4LS, UK

All rights reserved.
Printed in China.

A catalogue record for this book is available from the British Library.

HB ISBN: 978-1-80505-347-7
PB ISBN: 978-1-80505-392-7

Written by:
Noah Leatherland

Edited by:
Rebecca Phillips-Bartlett

Designed by:
Ker Ker Lee

All facts, statistics, web addresses and URLs in this book were verified as valid and accurate at time of writing. No responsibility for any changes to external websites or references can be accepted by either the author or publisher.

FSC MIX Paper from responsible sources FSC® C113515

PHOTO CREDITS: All images courtesy of Shutterstock. With thanks to Getty Images, Thinkstock Photo and iStockphoto.
Recurring images: Alexander_Evgenyevich, Baranovska Oksana, holaillustrations, Mia S PARK, Nesterova. Cover — Mark Brandon, Tonographer. 2–3 — Julian Wiskemanni. 4–5 — Stefan Neumann, Zack Frank. 6–7 — haveseen, Seth Yarkony. 8–9 — kzww, Matt Filosa. 10–11 — Torsten Pursche, Makarenko Igor, Alfmaler. 12–13 —David R Butler, Lavendulan. 14–15 — Henrik Larsson, Macronatura.es, zdenek_macat, Danilashchyk Olena. 16–17 — Tony Northrup, Kalaeva, zulkamalober. 18–19 — Christopher Seufert, Stewart Kirk, pisanstock. 20–21 — Mike Korostelev, Patrick Messier, SL-Photography, mollyw. 22–23 — ThomBal, Markos Loizou.

CONTENTS

Page 4	In the Sand
Page 6	Hermit Crabs
Page 8	Lugworms
Page 10	Sea Lions
Page 12	Sand Dollars
Page 14	Mini Beasts of the Beach
Page 16	Sea Turtles
Page 18	Clams
Page 20	Penguins
Page 22	Creatures of the Coast
Page 24	Glossary and Index

Words that look like this can be found in the glossary on page 24.

IN THE SAND

If you visit the coast, you might find a sandy beach. A lot of amazing creatures make their homes in the sand. Next time you are at the coast, these creatures might be there too.

The coast is where the land meets the sea.

Some creatures spend their whole lives in the sand. Others live out at sea and only come onto the sand at certain times of the year. Keep an eye out for them!

HERMIT CRABS

HERMIT CRABS ARE OMNIVORES. THEY EAT ANYTHING THEY CAN FIND.

Did you know that hermit crabs are not born with their shells? They search through the sand for a shell to make their home. Hermit crabs might even fight each other over a shell.

Hermit crabs test shells to make sure they can comfortably fit inside. Hermit crabs need to find bigger shells as they grow. They make their homes in many shells throughout their lives.

LUGWORMS

Have you ever come across curly piles of sand on the beach? These are made by lugworms. Lugworms make burrows to live in by swallowing sand and pooping it out.

Lugworms can be black, brown, pink or green. They can grow to around 40 centimetres long. Many coastal birds hunt lugworms by pulling them out of their underground burrows.

SEA LIONS

Sea lions spend a lot of time at sea, but sometimes they gather on beaches in large groups. They climb onto the sand when they are ready to have babies.

SEA LIONS ARE MAMMALS.

MANE

Sea lions use their flippers to crawl across the sand to find a place to rest. Sea lions got their name because they are large, they roar and some have manes just like lions do.

Sand Dollars

SOFT, HAIRY SPINES

Sand dollars are flat creatures. They have lots of soft spines and hairs covering their bodies. Sand dollars use these spines to move, eat and bury themselves in the sand.

Sand dollars usually live on sandy patches of the ocean floor. Sometimes, sand dollars can be found in the sand near the edge of the water. When they die, their skeletons often wash up onto beaches.

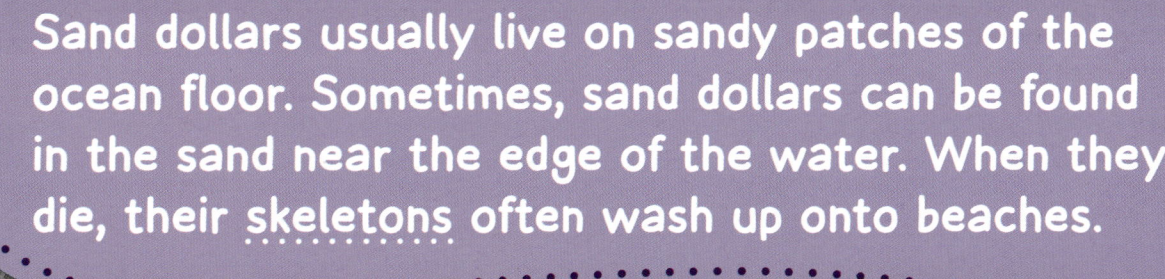

Some stories say that sand dollars are coins dropped by mermaids.

Mini Beasts of the Beach

KELP FLY

ROVE BEETLE

Lots of bugs live along the coast. The moist environment of the seaside makes a perfect home for bugs such as rove beetles. Kelp flies eat slimy seaweed found on beaches.

Some mini beasts have adapted to life in the sand. Beach wolf spiders are light brown and have spots on their bodies. This works as camouflage that helps them blend into the sand.

BEACH WOLF SPIDER

SEA TURTLES

SEA TURTLE EGGS

Sea turtles spend most of their lives in the ocean, but their lives begin in the sand. Mother turtles crawl onto the beach to bury their eggs in the sand.

The baby turtles hatch from the eggs and dig their way out of the sand. Then, they dash towards the sea. This can be very dangerous because of predators, such as birds and crabs.

CLAMS

Clams are soft creatures that live inside their shells. Some clams can be found in burrows on sandy beaches. Burying themselves in the sand protects them from predators, such as sea birds.

Clams open their shells to filter-feed.

Clams eat by filter-feeding. This means that they take in water from around them and eat tiny pieces of food found in the water. Then the water is pushed back out.

PENGUINS

People often think of penguins living in the Antarctic, but many types of penguins live on sandy beaches. African penguins live on sandy beaches in groups called colonies.

In their colonies, African penguins groom each other and work together to make burrows in the sand. These burrows are safe places to lay their eggs. African penguins are very noisy birds. They make sounds like donkeys.

CREATURES OF THE COAST

Sandy coasts are important to many creatures all around the world. Some animals spend their whole lives in the sand. Other creatures come to the beach to have babies and raise their young.

From mini beasts to massive sea lions, all sorts of creatures can be seen on the sand. Next time you go to the seaside, see what animals you can spot.

GLOSSARY

adapted	changed over time to suit the environment
burrows	homes made by animals by digging into the ground
camouflage	blend into the surroundings or background
groom	to brush and clean an animal's coat or fur
mammals	animals that are warm-blooded, have backbones and produce milk
moist	slightly wet
omnivores	animals that eats both plants and other animals
predators	animals that hunt other animals for food
skeletons	the framework of bones supporting the body

INDEX

burrows 8-9, 18, 21
eggs 16-17, 21
flippers 11
predators 17-18

seaweed 14
shells 6-7, 18-19
skeletons 13